SCIENCE FACTORY

WATER
& BOATS

JON RICHARDS

PowerKiDS
press

New York

Published in 2008 by The Rosen Publishing Group, Inc.
29 East 21st Street, New York, NY 10010

Design:
David West Books

Designer:
Flick Killerby

Illustrator:
Ian Moores

Consultant:
Steve Parker

Photographer:
By Roger Vlitos

Library of Congress Cataloging-in-Publication Data

Richards, Jon, 1970–
Water & boats / Jon Richards.
p. cm. — (Science factory)
Includes index.
ISBN-13: 978-1-4042-3909-8 (library binding)
ISBN-10: 1-4042-3909-X (library binding)
1. Water—Juvenile literature. 2. Water—Experiments—Juvenile literature. I. Title. II. Title: Water and boats.
GB662.3.R53 2008
532.078—dc22
2007016692

Manufactured in the United States of America

INTRODUCTION

Water is amazing stuff! It comes in three different forms: as a solid, as a liquid, and as a gas. Some objects can float in it, other objects can sink in it, and a few objects can both float and sink in it! Water can be used to turn wheels, and it can even climb dozens of feet (m) against the force of gravity! Read on and discover a whole host of experiments, as well as the why-it-works boxes that will teach you more about water.

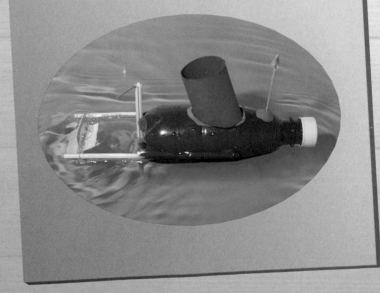

CONTENTS

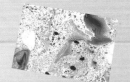

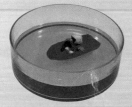

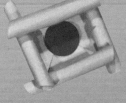

YOUR FACTORY

BEFORE YOU START any of the experiments, it is important that you learn a few simple rules about the care of your science factory.

● Always keep your hands and the work surfaces clean. Dirt can damage results and ruin an experiment!

● Read the instructions carefully before you start each experiment.

● Make sure you have all the equipment you need for the experiment (see checklist opposite).

● If you don't have the right piece of equipment, then improvise. For example, a dishwashing-liquid bottle will work just as well as a plastic drink bottle.

● Don't be afraid to make mistakes. Just start again — patience is very important!

Equipment checklist:
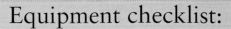

- Scissors, tape, and glue
- Ice-cube tray
- Plastic bottles and fruit-juice cartons
- Oil-based paints and turpentine
- Potato
- Salt and food coloring
- Paper, cardboard, and tissue paper
- Modeling clay
- Large bowl
- Aluminum foil and colored cellophane
- Paper clips and wire
- Dishwashing liquid
- Drinking straws and balloons
- Plastic cups and trays
- Toothpicks, sticks, and matchsticks
- Cotton thread
- Rubber bands and plastic food wrap
- Pipe cleaners and bottle caps

WARNING:

Some of the experiments in this book need the help of an adult. Always ask a grown-up for help when you are using scissors or electrical objects, such as hair dryers!

ICE AND WATER

IF LIQUID WATER IS COOLED ENOUGH, it turns into a solid lump. This solid form of water is called ice. Unlike most substances that shrink when they freeze, water gets larger, or expands, when it turns into ice. Because it expands, it becomes less dense, or lighter, than liquid water. This is why ice cubes float in a drink. This experiment will let you examine the different forms of water, and show you what happens when water freezes and when it melts.

WHAT YOU NEED
Plastic bottle
Food coloring
Ice-cube tray

WHY IT WORKS

As the ice melts, it turns into water. Because this melted ice is cooler than the warm water, it is also denser. As a result, this freshly melted ice sinks. As it sinks it warms up, becomes less dense, and so rises again.

MELTING ICE WATER SINKS, IS WARMED, AND SO RISES IN A CIRCULAR CURRENT.

MELTING ICEBERG

1 Carefully cut the top off a clear plastic drink bottle. Run warm water from the faucet into the bottle. Add food coloring and stir well.

2 Fill another bottle with cold water. Add a few drops of a different food coloring and stir the mixture well. Pour the water into an ice-cube tray and leave it in the freezer overnight.

5 *Pour the salty water into a clear plastic bottle from which the top has been cut off. Slowly add the colored water by pouring it over the back of a spoon. Place your fish carefully on the water's surface. Watch the fish sink through the colored water, but float on top of the salty water.*

WHY IT WORKS

The salt water is denser than the colored water, so it will stay at the bottom of the bottle, below the colored freshwater. Your fish is denser, or heavier, than the colored freshwater, but less dense than the salt water. As a result, it will float between the two layers of water.

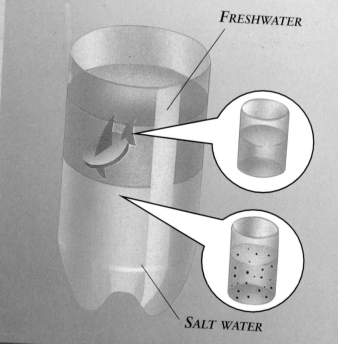

FRESHWATER

SALT WATER

OIL SLICK

WHAT YOU NEED
Oil-based paints
Turpentine
Paper
Large bowl

JUST AS ICE FLOATS IN WATER BECAUSE IT IS LIGHTER, or less dense, some liquids will float on the surface of water as well, because they are also less dense. You may have noticed a film of oil floating on top of a puddle on a rainy day. This experiment lets you use floating oil paints to make patterns on paper.

You can make a swirling, multicolored pattern with oil-based paints and water.

10

SWIRLING PATTERNS

1 Mix some different colored oil-based paints with turpentine to make them thinner.

2 Fill a plastic bowl with cold water and carefully pour small amounts of the paints onto the surface.

3 Gently swirl the paints around using a clean stick.

4 Carefully lower a sheet of paper on top of the paint. Allow the paper to soak up some of the paint, then peel off the paper and see what patterns have been left on it by the paint.

WHY IT WORKS

The paint floats on the surface of the water because it is less dense than the water. As a result, you can soak up the colors by placing your paper on top.

OIL FLOATS ON TOP OF WATER BECAUSE IT IS LESS DENSE.

SALT WATER LIES BENEATH FRESHWATER BECAUSE IT IS DENSER.

FLOATING STRAWS

A hydrometer is a device that measures density. You can make one using a drinking straw and some modeling clay. The hydrometer will float higher in dense liquids than in less-dense liquids.

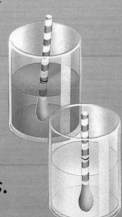

SHIPSHAPE

ENORMOUS AIRCRAFT CARRIERS AND LARGE CRUISE SHIPS float, yet a single metal screw will sink! When it comes to floating, size is not important. Instead, whether something floats or not depends on the weight of the water displaced by the object when it sits in the water. If the displaced water weighs more than the object, then the object will float.

CRAFTY VESSELS

1 Mold a lump of modeling clay into different solid shapes and see if they will float in a bowl of water.

2 Now roll the clay flat. Curve the edges up and pinch them together to form a boat shape. Make sure your boat doesn't leak!

3 Gently place your boat into the bowl of water and see if it floats. Mark on the side of your boat the level that the water reaches.

4 Now make a clay figure to sit in the middle of your boat. Put the boat back in the water and you will see that the boat now sits lower in the water than when it was empty.

WHY IT WORKS

When you shape your boat, the volume of water displaced by the boat weighs more than the boat, so it floats. When you add your clay passenger, you are increasing the weight of the boat, so it sinks slightly into the water.

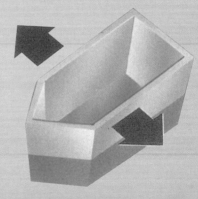

FLOAT OR SINK?

Try making different boat shapes. A high-sided boat will float better than a shallow one. This is because it can sit lower in the water without having water spill over its sides. Now try some other boat shapes, and see which will carry the heaviest load.

HIGH-SIDED BOAT

WATER FLOWS OVER SIDES OF BOAT

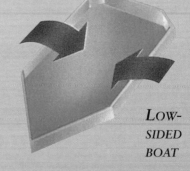

LOW-SIDED BOAT

REACHING THE DEPTHS

UNLIKE BOATS, SUBMARINES CAN SINK AND FLOAT as many times as they want. They do this by pumping air or water into special tanks inside them. If a submarine pumps water into these tanks, then it becomes heavier and sinks. If it pumps air into these tanks, then the submarine becomes lighter and rises toward the surface. Build your own submersible and see how it can sink and float as many times as you like!

WHAT YOU NEED
Drinking straw
Paper clip
Modeling clay
Plastic bottle

Squeeze the sides of the bottle and watch your submersible dive to the bottom. Release the bottle and your submersible will rise to the top.

BOTTLED SUBMERSIBLE

1 To make a submersible, cut the ribbed part off a flexible drinking straw and bend it in half.

2 Open up a metal paper clip and push each end into each end of the bent straw. Make sure that the paper clip will not slide out.

3 Roll out three thin strips of modeling clay. Loop and pinch each one around the paper clip. These strips will weigh your submersible down.

4 Place your submersible in a glass of water to test that it floats the right way up. Alter the amount of modeling clay until it floats.

WHY IT WORKS

Trapped inside the straw is a bubble of air. When you squeeze the bottle, water is pushed into the straw and squashes the air bubble. As a result, your submersible becomes heavier and sinks.

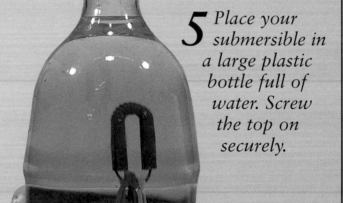

5 Place your submersible in a large plastic bottle full of water. Screw the top on securely.

UNDER PRESSURE

Stretch plastic wrap over the top of a tub of water and secure it with a rubber band. Push on the plastic. Does your submersible still sink?

CLIMBING WATER

HAVE YOU EVER WONDERED how plants can get water to every branch, stem, and leaf? This is due to a process called capillary action. It involves very long and very thin tubes that lie inside the plant. Forces inside these very narrow tubes draw water up. In the tallest trees, water can be pulled up dozens of feet (m)!

WHAT YOU NEED
Thin cardboard
Large bowl
Glue

CHANGING COLOR

Put some colored water in a vase with a flower. Over a couple of days, the flower will draw the colored water up its stem and its flower will change color.

PAPER PETALS

1 Take a square piece of smooth writing paper or cardboard and fold it in half. Do not use shiny paper.

2 Now fold the paper in half again to form a square.

3 Fold it in half again, this time to form a triangle.

7 Place your flowers in a bowl of water. Watch as the petals of your flowers unfurl as the water seeps into the paper.

4 Cut the shape of a petal out of the side with the thickest fold. Unfold the paper and you will have your flower.

5 Using a pencil or a straw, roll the petals of the flower so that the petals remain closed.

6 Brighten up your flower by sticking a circle of different colored paper in the center. Make some other flowers using different shapes and colors.

WHY IT WORKS

Like the stem of a plant, the paper is full of tiny tubes called capillaries. An attractive force between the molecules of water and the sides of these tiny tubes is strong enough to draw water up. As the water rises, the paper becomes heavy and the petals unfurl.

WATER LEVEL

CAPILLARY TUBES IN PAPER

WALKING ON WATER

IF YOU SPILL SOME WATER ONTO A FLAT SURFACE, you will see that the drops of water will clump together — almost as if they are held together by an invisible skin. You can also see this if you fill a tall glass with water right to the very top. The water will appear to bulge slightly above the top of the glass. Some insects use this invisible skin and can actually walk on water!

WATER WALKERS

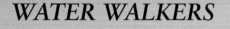

1 *Use some paper clips and wide strips of foil to make your water walkers.*

2 *Wrap each paper clip in a strip of foil and twist three legs out of the foil on either side of your water walker. Color your water walkers with bright colors.*

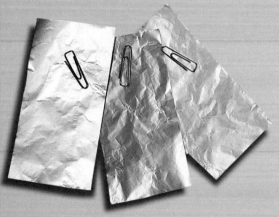

3 *Place your water walkers on a sheet of tissue paper. Hold the tissue tightly and gently lower it onto the surface of a bowl of water.*

4 *The tissue will gradually soak up the water and sink to the bottom, but your water walkers will rest on the surface.*

WHY IT WORKS

Water molecules are attracted to each other. On the surface, this attraction pulls the molecules together, producing a force called surface tension. This force is strong enough to support some light objects.

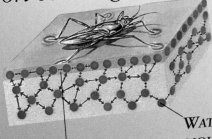

SURFACE TENSION BETWEEN WATER MOLECULES

WATER MOLECULES

MOVING INSECTS

Liquid soap breaks up the water's surface tension. By dropping some soap between your water walkers, you will make them dart away from each other.

19

THE POWER OF WATER

UNLIKE AIR, LIQUIDS CANNOT BE SQUASHED. This makes liquids very useful in lifting heavy objects, from raising the bed of a dump truck to stretching the arm of a large digger. This use of liquids to lift and move objects is called hydraulics. Build your own hydraulic machine that will show you how water can be used to lift loads.

LIQUID LIFT

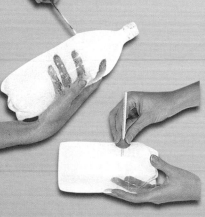

1 Carefully cut the tops off two plastic bottles and make them the same height.

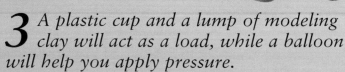

2 Pierce a hole in the side of each bottle a short distance up from the bottom. Push a drinking straw through the holes to link the bottles. Use modeling clay to seal the joints and make them watertight.

3 A plastic cup and a lump of modeling clay will act as a load, while a balloon will help you apply pressure.

4 Color some water with food coloring and fill the bottles so that they are about two-thirds full. Float the cup in one bottle and place the lump of modeling clay in the cup.

WHY IT WORKS

When you push down on the balloon, it forces water through the straw, transferring the force of your push into the other bottle. More water now sits in the other bottle, so the cup sits higher than it used to.

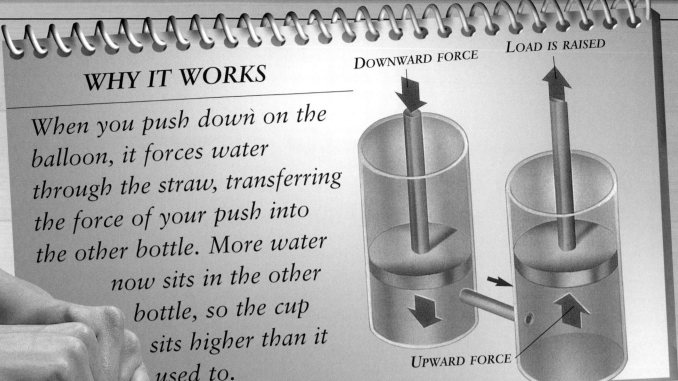

DOWNWARD FORCE

LOAD IS RAISED

UPWARD FORCE

5 Inflate the balloon a little, and place it in the other bottle. The balloon must press and slide against the bottle's sides. Now push down on the balloon, and watch the cup and the modeling clay rise in the other bottle.

CHANGING THE SYSTEM

Try using bottles of different sizes. You will find that a tall, narrow bottle will raise the load the highest because the displaced water has to fit into a narrower bottle.

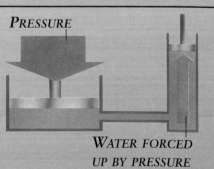

PRESSURE

WATER FORCED
UP BY PRESSURE

WATER JETS

BECAUSE OF GRAVITY, WATER USUALLY FLOWS DOWNHILL — however, it can be made to shoot up into the air! These jets of water can be natural, such as geysers, or artificial, such as fountains. This experiment lets you build your own fountain. See how high you can make your jet of water soar up into the air.

WHAT YOU NEED
Plastic bottle
Drinking straws
Modeling clay
Plastic tray
Large bowl

LIQUID JETS

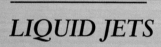

1 Carefully cut the bottom off a large plastic bottle.

2 Seal the mouth of the bottle with some modeling clay and poke a straw through into the bottle. Fit another straw to the end of the first straw to form a U-shaped bend.

3 Pierce two holes in the bottom of a plastic tray. Turn the bottle upside down and feed the straw tube through the holes. Seal the bottle in position over one of the holes using modeling clay.

A FALL IN PRESSURE

Pierce holes down the side of a plastic bottle. Fill the bottle with water and watch how far the jets of water squirt. The jets at the bottom, where the pressure is greatest, will squirt farther than those at the top.

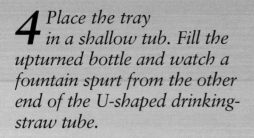

4 *Place the tray in a shallow tub. Fill the upturned bottle and watch a fountain spurt from the other end of the U-shaped drinking-straw tube.*

WHY IT WORKS

The fountain is caused by the weight of the water in the upturned bottle. This weight causes a build-up of pressure. The more water there is in the bottle, the greater the pressure, and the higher your fountain will soar. This pressure is released when the water squirts out of the U-shaped bend.

WATER IN BOTTLE
CAUSES PRESSURE

PRESSURE OF WATER
CAUSES FOUNTAIN

WATERWORKS

WATER HAS BEEN USED as a source of power for thousands of years. Since Roman times, wooden waterwheels have powered millstones to grind corn into flour. Today, enormous dams channel fast-flowing water past the modern version of the wooden waterwheel — the turbine. This spins to make electricity.

Place the waterwheel in a bowl. Pour water into the upside-down bottle. Watch as water pours onto your waterwheel, causing it to spin and raise the bucket.

WHAT YOU NEED
Plastic bottles
Drinking straw
Tape
Toothpicks
Matchstick
Stick
Cotton thread
Large bowl
Bottle cap

WHY IT WORKS

The waterwheel uses the energy of the falling water to make it spin. As it spins, the wheel winds up the thread, raising the bucket.

FALLING WATER
SPINS THE WHEEL

WATER POWER

1 Cut the bottom off a liquid soap bottle to make your waterwheel. Cut out four flaps from the side of the wheel. Bend these flaps as shown to make the wheel's blades. Make a hole in the center of the wheel.

2 Cut a section out of the bottom of a plastic bottle, large enough for the wheel to fit into. Pierce holes on either side of this section.

3 Fit the waterwheel into the cutout section by passing a drinking straw through the holes in the bottle and in the wheel. Attach the wheel to the straw using modeling clay. Poke toothpicks through the ends of the straw to hold it in place.

4 Pierce holes in the top of the plastic bottle and feed a stick through. Tape a piece of drinking straw to one end of the stick.

5 Make a bucket using a bottle cap. Glue a matchstick across the top of the cap and tie a piece of cotton thread to the matchstick. Feed the thread through the short straw and tie it around a toothpick pushed through the straw holding the waterwheel.

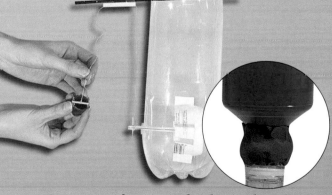

6 Fix another upside-down bottle on top, having cut off its bottom and sealed its top, leaving only a small hole.

GETTING HEAVY

Try adding small modeling-clay weights to the bucket. See how this affects the speed at which the bucket is lifted. You will find that with more weight to lift, the wheel will find it harder to raise the bucket. You could also try raising the height of the upside-down bottle — how does this affect the wheel?

PADDLING AWAY

WHAT YOU NEED
Small plastic bottle
Cardboard
Two sticks
Fruit-juice carton
Modeling clay
Rubber band

SOME OF THE EARLIEST POWERED BOATS WERE CALLED PADDLE STEAMERS. They used wheels, either at the rear of the boat or hung on its sides, to push the vessel through the water. However, the age of the paddle steamer was short-lived. Before long, boat builders found that propellers were better at pushing boats. Paddle steamers can still be seen today, but mostly as tourist attractions.

PADDLE POWER

1 *Screw the top of a small plastic drink bottle on tightly. Cut a hole in the side of the bottle where the boat's funnel will sit.*

2 *Tape two sticks to the sides of the bottle so that they stick out past the bottom of the bottle.*

3 *Cut two rectangles from a fruit-juice carton, making sure that they are not as wide as the plastic bottle.*

4 Make a slit halfway down each rectangle and slide the two together to form a cross-shaped paddle.

5 Tape a rubber band to the paddle and fix the ends of the band around the sticks. Make sure that the paddle does not touch the boat.

When you wind the rubber band, you are storing energy in it. This energy is released when you let go of the paddle, causing it to spin. As it spins, the blades push against the water, moving the boat forward.

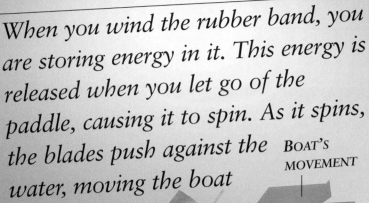

BOAT'S MOVEMENT

SPINNING PADDLE WHEEL

6 Weigh down the boat by placing a lump of modeling clay in the bottle. Cover the hole in the bottle with a cardboard funnel. Wind up the rubber band and place your boat in a tub of water. Watch as the paddle spins and the boat moves forward.

SIZE MATTERS

Try different-sized paddle wheels on your boat. You will find that a larger wheel will push the boat along more quickly than a small one.

27

STEERING

Fish use fins on their bodies to steer themselves through the water. Similarly, boats have one fin at the rear that is used to steer, called a rudder. In addition to a rudder, submarines have fins on their bodies, just like fish. They can use these fins to move the submarine up and down and from side to side as she travels underwater.

WHAT YOU
NEED
Wire
Drinking straw
Fruit-juice
carton
Paddleboat
made on pages
26-27

*By moving
your rudder from
side to side, you can make
your paddleboat alter its course.*

WHY IT WORKS

The rudder works by deflecting the water, causing the boat to alter its course. If the rudder points straight back, then the boat will go straight (1). If the rudder is turned to the right, then the boat will steer right (2) and left if the rudder is turned to the left (3).

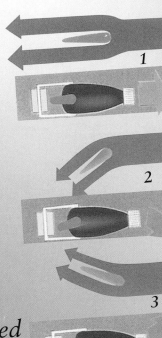

SUBMARINE STEERING

Make a submarine out of a lump of modeling clay. Fix four fins to the sides, two at the front and two at the back. Adjust these fins to point up or down. See how they affect your submarine's descent through a bottle of water.

SUBMARINE WITH FINS

RUDDER CONTROL

1 To make a rudder, bend a piece of wire to form a right angle, making the handle of your rudder.

2 Cut out a small square from a fruit-juice carton. Slide the wire through a piece of straw and attach the square to the bottom of the wire. Attach the straw to the back of the paddleboat made on pages 26-27.

3 Wind up the paddle and place the boat in a tub of water. As the boat moves forward, turn the handle of the rudder from one side to the other, and see what this does to your boat's course.

GLOSSARY

BOILING (BOY-ul-ing) The process of turning a liquid into a gas by raising its temperature. You can boil liquid water to turn it into steam. *Find out how you can boil salt water to separate the salt from the water on page 8.*

CAPILLARY ACTION (KA-puh-ler-ee AK-shun) The movement of a liquid along very narrow tubes. *You can see how capillary action works in the project on pages 16-17.*

DENSITY (DENT-suh-tee) The heaviness of a substance for a particular volume. *Find out about how the density of water changes depending on the form it is in on page 6.*

FRESHWATER (FRESH-wah-ter) Water that has very few dissolved substances, such as salts and minerals. *You can see how the density of freshwater affects whether or not objects float on pages 8-9.*

GRAVITY (GRA-vuh-tee) The attractive force between objects. The earth's gravity keeps us on the ground. *You can find out how to make water shoot up into the air and seem to go against gravity in the project on pages 22-23.*

HYDRAULICS (hy-DRO-liks) The technology of liquids. *You can see how a hydraulic machine can help you lift things in the project on pages 20-21.*

GLOSSARY

HYDROMETER (hy-DRAH-meh-ter) An instrument used to compare the densities of different liquids. *Find out how to make a hydrometer on page 11.*

MOLECULE (MAH-lih-kyool) The smallest naturally occurring particle of a substance. *See how salt molecules change water on pages 8-9.*

RUDDER (RUH-der) A special paddle usually found at the rear of a boat. It is used to steer the boat. *You can see how a rudder makes a boat move on pages 28-29.*

SUBMERSIBLE (sub-MER-suh-bul) A craft that can go underwater and rise again to the surface at will. *Find out how to build your own submersible on pages 14-15.*

SURFACE TENSION (SUR-fus TENT-shun) The attractive force between molecules at the surface of a liquid. *Experiment with surface tension on pages 18-19.*

WATERWHEEL (WAH-ter-hweel) A wheel that is turned by flowing water. Today, waterwheels, or turbines, are used to produce electricity. *Make a waterwheel in the project on pages 24-25.*

INDEX